CONTENTS

COMPLIANCE REVIEW PROCESS AND MISSILE DEFENSE

MONDAY, JULY 21, 1997

U.S. SENATE,
SUBCOMMITTEE ON INTERNATIONAL SECURITY,
PROLIFERATION AND FEDERAL SERVICES,
OF THE COMMITTEE ON GOVERNMENTAL AFFAIRS,
Washington, DC.

The Subcommittee met, pursuant to notice, at 2:30 p.m., in room SD–342, Dirksen Senate Office Building, Hon. Thad Cochran, Chairman of the Subcommittee, presiding.

Present: Senators Cochran and Levin.

OPENING STATEMENT OF SENATOR COCHRAN

Senator COCHRAN. The meeting for the Subcommittee will please come to order. Today we have as the subject of our hearing, "The Compliance Review Process and Missile Defense."

This past March in Helsinki, Presidents Clinton and Yelstin issued a joint statement announcing the outlines of an agreement that would draw a demarcation line between theater and strategic missile defenses, and they sent their negotiators back to Geneva to finalize an agreement. As has happened so many times in the past, however, the negotiators were unable to agree on the important details and returned last month empty-handed. But even if they had reached an agreement based on the Helsinki joint statement, it would not have drawn a clear demarcation line. Any agreement based upon the Helsinki statement, after all, offers clarity only for far less capable theater missile defenses over which there has been little dispute. For the more capable systems, which have been at the heart of the nearly 4-year-long ABM–TMD demarcation negotiations, Helsinki contains no agreement on demarcation, only an acknowledgment that the Treaty parties will continue to determine for themselves whether their more capable systems comply with the ABM Treaty.

The process by which the United States makes those determinations will continue to be critically important, not only for TMD systems, but also for National Missile Defense, and is the subject of our hearing today.

The U.S. Compliance Review Process has, in the past, produced results that are unfortunate. For example, early in 1994, the administration declared that testing the prototype THAAD interceptor would be non-compliant and that it would have to be treated as an ABM system, even though it could never be usefully employed as an ABM system. A year later, the prototype was determined to be compliant, provided the capabilities of the system were substantially reduced by removing its ability to receive satellite cuing data for its radar. According to the Administration, the effort required to modify the software in accordance with that decision cost the U.S. taxpayer several million dollars. Then the Administration announced in September of 1996 that not only could the prototype THAAD be tested, so could the final, objective system, and both could now be equipped with the same satellite cuing capability that we had paid to take out of the system earlier. These inconsistencies raise serious questions about how these reviews are conducted.

The Compliance Review Process also passes judgments on our National Missile Defense System, and compliance determinations will be increasingly important as we move forward on that program. Numerous Administration officials, including the Director of the Arms Control and Disarmament Agency in testimony before this Subcommittee in May of this year, have said that our NMD deployment program may have elements that do not comply with the ABM Treaty, and might require further negotiation with the Russians. For example, Article I of the ABM Treaty explicitly prohibits defense of the Nation's territory from strategic ballistic missile attack, yet that is precisely the purpose of our NMD program. There are numerous other potential features of an NMD system that seem to conflict with the ABM Treaty, as Director Holum of ACDA testified in our May hearing. Whether these apparent incompatibilities can be rationalized away will be determined by the same Compliance Review Process that governs our TMD programs.

We have as our witness at today's hearing someone who deals with these issues every day. Dr. Kent Stansberry Chairs the Defense Department's Compliance Review Group in his capacity as Deputy Director for Arms Control Implementation and Compliance in the Office of the Under Secretary of Defense for Acquisition and Technology. As Chairman of the Compliance Review Group, he is responsible for running the process that determines whether U.S. weapon systems comply with our treaty obligations. We are pleased to have him here to help us better understand the Compliance Review Process as it applies to missile defense.

Today's hearing may touch on matters that are classified and we are prepared to move to a closed session if that becomes necessary. Our questions will be unclassified, but if your answers require a classified response Dr. Stansberry, let us know and we will return to those specific questions in closed session.

I am happy to be joined by my distinguished colleague from the state of Michigan, Senator Carl Levin today and I yield to him at this point for any opening statement or comments that he would care to make. Senator Levin.

OPENING STATEMENT OF SENATOR LEVIN

Senator LEVIN. Thank you, Mr. Chairman. And with you, I want to join in welcoming our witness here today. This hearing is going to provide the Subcommittee and the Senate and the public with an opportunity to learn more about how the Executive Branch reviews U.S. military programs and systems for arms control compli-

S. Hrg. 105–243

COMPLIANCE REVIEW PROCESS AND MISSILE DEFENSE

HEARING

BEFORE THE

SUBCOMMITTEE ON INTERNATIONAL SECURITY, PROLIFERATION, AND FEDERAL SERVICES

OF THE

COMMITTEE ON GOVERNMENTAL AFFAIRS
UNITED STATES SENATE

ONE HUNDRED FIFTH CONGRESS

FIRST SESSION

JULY 21, 1997

Printed for the use of the Committee on Governmental Affairs

U.S. GOVERNMENT PRINTING OFFICE

42–600 cc WASHINGTON : 1997

For sale by the U.S. Government Printing Office
Superintendent of Documents, Congressional Sales Office, Washington, DC 20402
ISBN 0-16-055870-0

COMMITTEE ON GOVERNMENTAL AFFAIRS

FRED THOMPSON, Tennessee, *Chairman*

SUSAN M. COLLINS, Maine
SAM BROWNBACK, Kansas
PETE V. DOMENICI, New Mexico
THAD COCHRAN, Mississippi
DON NICKLES, Oklahoma
ARLEN SPECTER, Pennsylvania
BOB SMITH, New Hampshire
ROBERT F. BENNETT, Utah

JOHN GLENN, Ohio
CARL LEVIN, Michigan
JOSEPH I. LIEBERMAN, Connecticut
DANIEL K. AKAKA, Hawaii
RICHARD J. DURBIN, Illinois
ROBERT G. TORRICELLI, New Jersey
MAX CLELAND, Georgia

Hannah S. Sistare, *Staff Director and Counsel*
Leonard Weiss, *Minority Staff Director*
Michal Sue Prosser, *Chief Clerk*

SUBCOMMITTEE ON INTERNATIONAL SECURITY, PROLIFERATION AND FEDERAL SERVICES

THAD COCHRAN, Mississippi, *Chairman*

SUSAN M. COLLINS, Maine
PETE V. DOMENICI, New Mexico
DON NICKLES, Oklahoma
ARLEN SPECTER, Pennsylvania
BOB SMITH, New Hampshire

CARL LEVIN, Michigan
DANIEL K. AKAKA, Hawaii
RICHARD J. DURBIN, Illinois
ROBERT G. TORRICELLI, New Jersey
MAX CLELAND, Georgia

Mitchel B. Kugler, *Staff Director*
Linda Gustitus, *Minority Staff Director*
Julie Sander, *Chief Clerk*

ance, particularly with respect to missile defenses and the ABM Treaty. This is an important National Security topic that is not widely understood.

Success of Administrations from President Nixon onward have viewed the ABM Treaty as an important Treaty because it has contributed to strategic stability, prevented an arms race in defensive and space weapons systems, and permitted very significant reductions in nuclear arsenals that were built up during the cold war. Secretary of Defense Cohen and General Shalikashvili, Chairman of our Joint Chiefs, have stated clearly, these reductions in excess nuclear weapons are squarely in our security interests. Now others may disagree with the value of the ABM Treaty, but I think most of us believe that we should live up to our Treaty obligations, just as we expect others to live up to theirs, unless and until we or other parties withdraw from or abrogate a Treaty.

United States Compliance Review Process is a key element in ensuring that the United States is in compliance with our arms control obligations. It is not a new function. It was set in place during the early 70's at the beginning of the era of nuclear arms limitation and has been supported by every Administration from President Nixon to President Bush through to the current Administration.

In the context of missile defense systems in the ABM Treaty, the compliance review process is a contentious issue. The treaty has a provision, Article VI(a) that prohibits either side from giving non-ABM systems, ABM capabilities, and prohibits testing non-ABM systems in an ABM mode. Thus, we need to know what the demarcation is between theater missile defenses and strategic or National missile defenses. Congress has urged the President to negotiate this demarcation and has provided guidelines for the demarcation. If we can negotiate a demarcation agreement with Russia, it will remove these theater systems from the table as potentially contentious issues. The Administration is currently negotiating this issue in Geneva.

The Senate has a uniquely constitutional role in providing advice and consent to treaties. Consequently, we have a strong interest in matters relating to arms control treaty issues, including compliance. This hearing will help to inform the Senate on these important issues and I want to thank you, Mr. Chairman, for convening this hearing to help us to understand the issues more fully and to have a record which will be available to all of our colleagues.

Senator COCHRAN. Thank you very much, Senator Levin, for your comments and your participation in this series of hearings that we are having on the subjects that are related to this issue today.

Dr. Stansberry, we have a copy of your prepared statement for which we thank you and we will make that a part of the record in full. We encourage you to make whatever comments you think would be helpful to our Subcommittee at this time. You may proceed.

TESTIMONY OF DR. KENT STANSBERRY, DEPUTY ASSISTANT FOR ARMS CONTROL IMPLEMENTATION AND COMPLIANCE OF THE UNDER SECRETARY OF DEFENSE (ACQUISITION AND TECHNOLOGY)

Dr. STANSBERRY. Thank you, Mr. Chairman, Senator Levin, and staff. I appreciate the opportunity to appear before you and discuss the Arms Control Compliance Review Process.

I plan to focus primarily on the application of that process, the ABM Treaty, and particularly theater missile defenses (TMD), under the ABM Treaty.

The Department of Defense decides the treaty compliance questions through an established review process. It was created in connection with the SALT I agreements, which includes the ABM Treaty, back in 1972. The process is defined completely and more clearly in a DOD Directive which I believe the staff has a copy of. Under this process, the military services and other defense agencies must seek compliance approval before taking any action that might reasonably raise a question of compliance. The Under Secretary of Defense for Acquisition and Technology, who is my boss, has been authorized by the Secretary to ensure compliance of the Department of Defense and to provide guidance to the services and agencies in coordination with the DOD General Counsel, the Under Secretary of Defense for Policy, and the Chairman of the Joint Chiefs of Staff. In practice, most of the detailed work is carried out by a Compliance Review Group. I Chair that group for the ABM Treaty.

Let me just speak briefly to the details of the process. A formal compliance review generally begins when a service or defense agency brings forward a particular plan or program with proposed activity. That activity is then compared with our obligations to determine whether it is acceptable or whether there may be a need for modifying the activity. In doing that, we determine the obligations through a general pattern that includes reading the Treaty text and its associated documents, and examining the record of Senate approval to determine how the Executive Branch explained the Treaty to the Senate as a part of our Constitutional process of making treaties. We also examine the relevant practices of the parties to determine how the parties have behaved in the past with respect to particular obligations and we often examine the negotiating record to consider how the parties understood the text as they negotiated it.

This formal process of compliance review ends typically with guidance to the service or agency that initiated their request. The guidance is generally provided by the Under Secretary for Acquisition and Technology with the concurrence of representatives of the Under Secretary for Policy, the General Counsel, and the Joint Staff. One particular strength of that process is that the final decisions represent consensus of those organizations. And in fact, every single past compliance determination has been a consensus. There has not been a dissenting view in any past decision.

Occasionally, we will raise issues with other agencies of the Executive Branch, particularly the State Department, the Arms Control Agency and the National Security Council staff, when the issues involve matters that may effect them.

Now, let me talk specifically about theater missile defense for a little bit. The ABM Treaty does not explicitly limit theater missile defenses, but implicitly limits them through a provision which was intended to prevent upgrade. Let me read that: Article VI of the Treaty States in part, "To enhance assurance of the effectiveness of the limitations on ABM systems and their components provided by the Treaty, each party undertakes: (a) not to give missiles, launchers, or radars, other than ABM interceptor missiles, ABM launchers, or ABM radars, capabilities to counter strategic ballistic missiles or their elements in flight trajectory, and not to test them in an ABM mode."

The United States was the sponsor of that language back in the early 1970's, when the Treaty was negotiated, primarily because of our concern about widely deployed air defenses in the Soviet Union and the possibility that those air defenses may represent a defensive capability against our ICBMs and SLBMs. That prohibition on upgrading non-ABM systems and components in Article VI(a) defines two quite specific obligations: (1) not to give capabilities to counter strategic ballistic missiles, and (2) not to test in an ABM mode. For the purpose of evaluating the capabilities to counter, we must use hypothetical capabilities because we cannot test our theater missile defense systems against strategic ballistic missiles to determine actual capabilities. Thus, we have assessed these capabilities to counter on the basis of computer simulations of performance of theater missile defense systems to counter strategic ballistic missiles. For example, in our evaluation of the Theater High-Altitude Area Defense (THAAD) System, we used computer simulations of the hypothetical performance of THAAD against Russian ICBMs and Russian SLBMs.

In assessing these simulations, we look extensively at past practices of the parties. A key element of that past practice was a report that was provided to Congress in February of 1986 by the Executive Branch where we discussed U.S. views about Soviet activity regarding air defenses. A key part of that was concluding that in making these determinations to apply Article VI(a), they must be made taking into account the military significance of whatever ABM capability is present in these non-ABM systems.

Now, as a practical matter we have assessed this military significance on the basis of the simulated hypothetical performance of a single TMD interceptor missile and radar to intercept a single reentry vehicle from certain Russian ICBMs and SLBMs.

As I mentioned, there is a second obligation of Article VI(a), not to test in an ABM mode. The meaning of that Article is substantially clearer because in the mid and late 1970's, we negotiated an agreed statement that provided substantial definition of the term, "tested in ABM mode" as it applies in the Treaty.

The Department of Defense has a number of TMD programs in various stages of development and deployment. All of these TMD programs are going forward without ABM Treaty effects on the capabilities necessary for them to meet their requirements. As the Department has previously informed the Congress, we have certified the compliance of the Patriot, Navy Area Defense, and Navy Theater Wide Systems and the THAAD System—in the case of THAAD, both the User Operational Evaluation System and the objective system. We have certified those compliant, as all of those programs are currently planned. Thus, we have certified as Treaty compliant all existing TMD programs that have matured to the point where it is possible for us to assess their compliance. These programs are proceeding without ABM Treaty effects on their intended capabilities for theater missile defense.

If the programs change, we will of course review them for compliance purposes in the future and make any additional compliance assessments as necessary based on program changes.

Now, let me speak briefly about National Missile Defense. The administration's policy with respect to ABM Treaty compliance in National Missile Defense is somewhat subtle. The policy is that development and testing of a National Missile Defense system will comply with the existing treaty. However, the system that we would deploy would be determined by the threat we face if we were to decide to deploy. That threat would decide the nature of the system architecture and then if that architecture requires that we change the ABM Treaty, we are prepared to seek those changes.

With respect to National Missile Defense, the Department is currently in the process of selecting a lead system integrator which will be responsible for development and if necessary deployment of an integrated National Missile Defense system. Given the immature status of that system design and the testing plans at this moment, it is not possible to make any definitive compliance assessments regarding National Missile Defense development testing. However, since the Treaty already specifically allows ABM testing, fully capable testing, of ABM systems in a fixed land-based mode, we expect that we will be able to carry out our National Missile Defense development and testing consistent with the Treaty. As I have mentioned, we would be willing to modify the Treaty to proceed with deployment.

I appreciate the opportunity to appear before the Subcommittee and I am prepared to help answer any questions that you may have.

[The prepared statement of Dr. Stansberry follows:]

PREPARED STATEMENT OF KENT G. STANSBERRY

Mr. Chairman, Members of the Subcommittee, and Staff: I wish to thank you for the opportunity to discuss with you the DOD arms control compliance review process. I plan to address the application of that process to the evaluation of missile defense programs and their compliance with the Anti-Ballistic Missile (ABM) Treaty. I plan to focus particular attention on the DOD theater missile defense CFMD) programs. I understand that ABM Treaty compliance for these programs is a specific interest of the subcommittee.

The Department of Defense decides treaty compliance questions through an established compliance review process. This process was created in connection with thee SALT I agreements, which included the ABM Treaty, in 1972. (This process is defined in DOD Directive 2060.1) Under this process, the Military Services and Defense Agencies must seek compliance approval before taking any action that would reasonably raise a compliance issue. The Under Secretary of Defense (Acquisition and Technology) has been authorized by the Secretary to ensure compliance for the Department of Defense and to provide compliance guidance to the Services and Agencies in coordination with the DOD General Counsel, the Under Secretary of Defense (Policy), and the Chairman of the Joint Chiefs of Staff. In practice, most of the detailed work supporting the compliance review process is carried out by the DOD Compliance Review Group, which is made up of representatives from the organizations just mentioned. I chair that Compliance Review Group for the ABM Treaty.

The advent of the Strategic Defense Initiative in 1984 led to increased activity in the DOD compliance review process for the ABM Treaty. Prior to that time, our ballistic missile defense program consisted primarily of studies and technology development. Since then, however, the arms control compliance review process has approved more than 100 individual tests or complete programs as consistent with the ABM Treaty. In summary, we have been successful in assuring that programs achieve their objectives while remaining in compliance with the ABM Treaty.

General Approach to Compliance Review

We have found through long experience in evaluating compliance questions that we need specific and detailed information in order to make compliance decisions. This is so because these decisions generally depend on the detailed nature of both the activity to be undertaken and the performance of the particular devices involved. Said differently, we have not found it useful or appropriate to try to address treaty compliance-questions in the abstract, but rather we address them in the context of information about specific and detailed plans. This consideration is important as we deal with compliance questions for our TMD programs.

A formal compliance review generally begins with a Military Service or Defense Agency bringing forward a plan or program for proposed activity. That activity is compared to U.S. arms control obligations to determine whether it is acceptable or would need to be modified. In determining the obligations, we follow a general pattern to assure that our interpretations are accurate and consistent in the complicated and ambiguous circumstances which often apply. The text of the Treaty and its associated documents, such as agreed statements and common understandings, fundamentally define the obligations. We can occasionally clarify the obligations by examining the Senate approval record to determine how the Executive Branch explained the treaty as part of our constitutional process of making treaties. We can further clarify the obligations by examining the relevant practices of the treaty parties to determine how the particular text has been interpreted in the past. These past practices can involve both the actions and the statements of the parties. Finally, we can clarify the obligations by examining the treaty negotiating record to consider how the parties understood the meaning during the process of drafting the text.

The formal compliance review process ends in guidance to the initiating Military Service or Defense Agency. The guidance is generally provided by the Under Secretary of Defense (Acquisition and Technology) with the concurrence of representatives of the Under Secretary of Defense (Policy), the DOD General Counsel, and the Joint Staff One basic strength of the compliance process is that the final decisions represent consensus. In fact, every past formal compliance determination has been a consensus. Compliance issues are also occasionally considered in the interagency context with the State Department, the Arms Control and Disarmament Agency, and the National Security Council Staff

Theater Missile Defenses and the ABM Treaty

Although the ABM Treaty does not explicitly limit TMD, it does so implicitly. Article VI of the ABM Treaty states, in part:

> To enhance assurance of the effectiveness of the limitations on ABM systems and their components provided by the Treaty, each Party undertakes: (a) not to give missiles, launchers, or radars, other than ABM interceptor missiles, ABM launchers, or ABM radars, capabilities to counter strategic ballistic missiles or their elements in flight trajectory, and not to test them in an ABM mode; . . .

This provision supports other Treaty limitations on the development, testing, and deployment of ABM capability by prohibiting "upgrading" non-ABM components to give them ABM capabilities.

When the ABM Treaty was negotiated, the issues associated with Article VI(a) were reasonably straightforward. At that time, the United States was concerned about the potential ABM capability of the widely deployed Soviet air defense systems and, thus, pressed during the negotiations for the inclusion of a prohibition on upgrading non-ABM components. More recently, however, particularly with the development of highly capable TMD systems, the issues associated with Article VI(a) have become more complex.

The prohibition on upgrading non-ABM systems and components in Article VI(a) of the ABM Treaty defines two specific obligations: (1) not to give "capabilities to counter strategic ballistic missiles" and (2) not to "test in an ABM mode." For the purpose of evaluating compliance of TMD systems with the first obligation, we assess hypothetical "capabilities to counter." They must be hypothetical capabilities because we cannot test TMD systems against strategic ballistic missiles to determine actual capabilities. Thus, we have assessed those capabilities on the basis of the simulated performance of TMD systems to counter strategic ballistic missiles. For example, in our evaluation of the compliance of the Theater High-Altitude Area Defense (THAAD) System, we used computer simulations of the hypothetical THAAD performance against Russian ICBMs and SLBMs.

In assessing the results of these simulations, we look extensively at the past practices of the ABM Treaty Parties. In February 1986, the Executive Branch, in a report to Congress submitted by the Director of the U.S. Arms Control and Disarmament Agency, discussed past Soviet air defense practices and reached the following conclusions:

> Clearly, the phrase "capabilities to counter" as used in the ABM Treaty was intended to have the ordinary meaning of "blocking" or "stopping" a reentry vehicle. In the context of the ABM Treaty giving a [surface-to-air missile system] "capabilities to counter strategic ballistic missiles" meant giving them actual ABM capabilities.
>
> Since virtually any air defense missile system has some level of ABM capability, the Treaty was not intended to preclude an incidental or insignificant ABM capability. Such a determination must ultimately be a factual determination taking into account the military significance of whatever ABM capability is present.

As a practical matter, we have assessed this "military significance" on the basis of the simulated, hypothetical performance of a single TMD interceptor missile and radar to intercept a single reentry vehicle from certain Russian (formerly Soviet) ICBMs or SLBMs.

The meaning of the second obligation of Article VI(a), not to "test in an ABM mode," is substantially clearer than the meaning of the first. The United States and the Soviet Union held discussions in the Standing Consultative Commission (the organization established by the ABM Treaty to promote the Treaty's objectives and implementation) during the 1970's on questions related to air defense activities under the Treaty. Those discussions resulted in an Agreed Statement of November 1978 giving a detailed definition of "tested in an ABM mode" as used in the Treaty.

The Department of Defense has a number of TMD programs in various stages of development and deployment. All of these TMD programs are going forward without ABM Treaty effects on the capabilities necessary to meet their requirements. As the Department has previously informed the Congress, we have certified the compliance of the Patriot, Navy Area Defense, and Navy Theater Wide systems and the THAAD system—both the User Operational Evaluation System and the objective versions of THAAD—as they are currently planned. DOD has now certified as treaty compliant all existing U.S. TMD programs that have matured to a point where it is possible to assess compliance. These programs are proceeding without any ABM Treaty effects on their intended capabilities for theater missile defense. We will of course, review for compliance purposes any changes to these programs. DOD will make any further compliance assessments that are necessary for U.S. TMD systems at the appropriate points in their programs.

The United States has been negotiating with Russia and other newly independent states to reach agreement on demarcation between ABM systems and non-ABM TMD systems. We have not yet concluded an agreement, and the negotiations are continuing. In the meantime, compliance determinations for TMD systems will remain a national responsibility. In light of the fact that we have already certified as treaty compliant all existing U.S. TMD programs that are sufficiently mature to allow a compliance assessment, we do not need the demarcation agreement in order for our TMD programs to proceed as currently planned.

National Missile Defense and the ABM Treaty

The DOD program for National Missile Defense (NMD) is currently in development. The program will be prepared, within three years, to deploy a NMD system within an additional three years, should a ballistic missile threat to the United States warrant such deployment. The Administration's policy is that development and testing in the NMD program will comply with the existing ABM Treaty. Should we decide to deploy, the system which we would deploy would be determined by the threat and might comply with the existing Treaty, or might require Treaty modification.

The Department is currently in the process of selecting a Lead System Integrator which will be responsible for developing and, if necessary, deploying an integrated NMD system. Given the immature status of the system design and testing plans, it is not possible at this time to make any definite compliance assessments regarding NMD development and testing. Since the ABM Treaty specifically allows development and testing of fixed, land-based ABM systems and components, we expect that we can carry out NMD development and testing consistent with the Treaty.

Thank you for the opportunity to appear before the subcommittee. I would be happy to answer any questions you may have.

Senator COCHRAN. Thank you very much, Dr. Stansberry.

You may have heard in my opening statement that I made a comment about inconsistencies that have raised questions about how the reviews under the current process are conducted. Specifically, I referred to the decision that was made in 1994 about the testing of the prototype THAAD interceptor. That it would be noncompliant and it would have to be treated as an ABM system, even though it could never be usefully employed as an ABM system, and then later the prototype was determined to be compliant provided that capabilities of the system were substantially reduced by removing its ability to receive satellite cueing data for its radar.

According to our information, the effort required a modification of software with that decision costing U.S. taxpayers several million dollars. And then the administration in 1996 announced that not only could the prototype THAAD be tested, so could the final objective system and both could now be equipped with the same satellite cuing capability that we had paid to take out of the system earlier.

Do you recall the circumstances surrounding those decisions and can you explain to us what happened in that situation and why?

Dr. STANSBERRY. I do recall. Some of that occurred while I was in my current position. Some of it preceded me. It does seem inconsistent on its face and I hope I can help clarify. The Administration, as a part of a process of reviewing the ABM Treaty when the Administration first took office, examined some of these questions. That review ended in the decision to proceed with negotiations on demarcation. Part of that review also examined what was then understood of THAAD's performance capability and concluded that THAAD, as then we understood it, would have to be treated as an ABM system under the ABM Treaty.

In January 1995 the THAAD Demonstration and Validation Program, basically the initial test program for THAAD, was approved on the basis that THAAD would not have some particular software to allow it to process information from external sensors to allow it to help perform its mission. That software was not needed for the Demonstration and Validation Program and it was on the basis that it was not going to be either needed or procured for that program that the approval was given.

As the program proceeded, it got to the point where it either needed to proceed with development of that software, or risk delays in the program. It was at that point that we did a much more substantial review of the performance consistent with the compliance review process I described earlier. Several things were different between 1993 and 1994 when the first reviews were carried out and in 1996 when we finished. There were several changes in the projected THAAD performance which were different. If the Subcommittee is interested, I can get into those changes in detail in closed session, but I am not prepared in open session.

42-600 - 97 - 2

There are several changes in the details of the performance. In one area we made some assumptions about the performance, assumptions which were conservative in the context of that analysis. It was conservative in the sense that it would err on the side of having the U.S. not do something illegal under the Treaty. Later, by September 1996, we understood the performance of THAAD much better and we understood the application of some of the particular assumptions much better so that with no changes to THAAD based on Treaty reasons, and no changes to the compliance review process, we were able with a much more detailed evaluation of THAAD's performance, we were able to conclude that it would have no ABM capability. It was on that basis that it was approved back in September.

Senator COCHRAN. You mentioned in your statement that one of the centerpiece considerations in the review process is the military significance standard. My staff tells me that in a recent briefing, Mr. O. J. Sheaks, who is Acting Assistant Director for Intelligence, and Verification Information Management at the Arms Control and Disarmament Agency, said that in judging compliance of Russian systems, their evaluations are "partly technical, partly political, and partly military significance."

Is military significance the standard used in assessing compliance of U.S. systems?

Dr. STANSBERRY. Mr. Chairman, I believe we also take into account the other two elements—the technical and the political. As I described in my statement—the simulated hypothetical performance of TMD systems against Russian strategic missiles is the technical element. The compliance review process, as I mentioned, involves participation from the Policy part of OSD and the General Counsel part of OSD. Part of the reason for their participation is to try to make sure that we maintain consistency between the way we are applying the treaty to the other side and the way we apply the treaty to ourselves.

Senator COCHRAN. In your assessments of the military significance, do you take into account operational factors, including basing modes, the number of launch platforms and missiles, the location of TMD systems and the character of a likely strategic attack?

Dr. STANSBERRY. To date, we have not. We have done that assessment on the basis of a single TMD system to counter a single incoming strategic reentry vehicle. We have been using that approach, as I mentioned earlier, to approve the compliance of all existing TMD systems. So, similar to many elements of law, we have gotten to the point in reviewing the practices of the parties and the meaning of the Treaty, that we needed to assess compliance and have simply left undecided further questions that are not necessary to make individual decisions.

Senator COCHRAN. We have been advised that the Arms Control and Disarmament Agency relies upon the Joint Staff in judging whether Russian TMD capabilities are militarily significant.

Do we also rely on the Joint Staff to judge U.S. capabilities in terms of their military significance? If we do not rely on the Joint Staff or involve them in any way, why do we not do that?

Dr. STANSBERRY. The Joint Staff is a part of the compliance review process as directed by the Secretary in the DOD Directive.

The Joint Staff, you probably need to talk to them to understand in more detail their view about this, but I believe, again, one of the things that they are sensitive to is making sure that we apply the Treaty to ourselves in a way that is consistent with the way that we apply it to other parties.

Senator COCHRAN. It would not be fair, then, to say that we are judging the Treaty against ourselves more narrowly than we judge it against possible Russian compliance with the treaty. Would that be fair or would it not be?

Dr. STANSBERRY. Well, in the end we have not found—let me back up. In the sense that we have allowed all existing programs to proceed, we have not found it necessary to identify some particular outer boundary—the outer boundary of compliance.

Senator COCHRAN. But it seems that there have been modifications and changes in plans and programs because of constraints imposed by those who are judging compliance with ABM provisions, is that not correct?

Dr. STANSBERRY. The THAAD program was not allowed to develop cueing software for more than a year based on compliance considerations. However, with the approval of THAAD last September, they were free to proceed with the development of that software without any restrictions on the way they planned to develop the program.

Senator COCHRAN. Thank you. Senator Levin.

Senator LEVIN. Was there a change in the software between the time it was not approved and the time it was approved?

Dr. STANSBERRY. I think in minor detail there was a change in specific requirements, but no change that was important to this particular issue. The software was allowed to proceed without any restrictions.

Senator LEVIN. What had changed then? Was it an understanding of what the software did or was it an interpretation of the Treaty requirements or something else that I can't think of?

Dr. STANSBERRY. Two general areas: The system itself changed in technical ways.

Senator LEVIN. You are saying those are unrelated to these considerations that we are talking about?

Dr. STANSBERRY. That is right. There were changes because of programmatic considerations.

Senator LEVIN. So then my question then remains was it approved relative to Treaty compliance a year later because of a different interpretation of where the demarcation line is or where the prohibitions were in the treaty in some other area? What changed in that year? I am not talking about the technical changes which are unrelated to the issue we are talking about. What changed relative——

Dr. STANSBERRY. Relative to the interpretation of the Treaty, nothing.

Senator LEVIN. What did change?

Dr. STANSBERRY. The capabilities intended to be in THAAD and some capabilities that were originally assumed to be in THAAD that we later found out were not going to be there. I would be prepared in closed session to discuss that in whatever detail that you are interested in.

Senator LEVIN. No, I do not for this session need the details as to what those changes were specifically. But it was the understanding then of what the capability of that system was that changed during the year?

Dr. STANSBERRY. That is correct, not the interpretation of Article VI(a) of the Treaty.

Senator LEVIN. Does the Department of Defense have the sole responsibility for compliance review of U.S. military systems and activities? I know there are consultations with the State Department or the Arms Control and Disarmament Agency, but is it the DOD's responsibility ultimately for that compliance review of military systems?

Dr. STANSBERRY. In a strict sense, yes. The Secretary is responsible for making sure that the activities of the Department are legal. He has designated the Under Secretary for Acquisition of Technology the responsibility to make sure that the Department's activities are consistent with arms control obligations, and then he has laid out a process for review of things within the Department. Occasionally when there are issues that we know are of concern to other agencies, we will go to them and discuss these issues with them as well.

Senator LEVIN. Now, does the Compliance Review Group for instance on the ABM Treaty, obtain the advice and the expertise of the Joint Staff and the JCS?

Dr. STANSBERRY. Yes. The Joint Staff participates. They are a member of the Compliance Review Group.

Senator LEVIN. What about the JCS, do they get involved or is their advice sought?

Dr. STANSBERRY. Do you mean the Chairman?

Senator LEVIN. Themselves.

Dr. STANSBERRY. The Joint Chiefs themselves?

Senator LEVIN. Right.

Dr. STANSBERRY. Occasionally issues of particularly sensitive issues will go all the way to the Chairman for his approval; and while I do not have a lot of insight into the details of the way the Joint Staff runs its process, I believe that their approval always represent the views of the Chairman and the Chiefs.

Senator LEVIN. So that these compliance reviews are not a matter of just a few attorneys sitting around theorizing as to what might or might not be compliant, you get the inputs here of the Joint Staff and the JCS?

Dr. STANSBERRY. And the Joint Staff coordinates on the decision.

Senator LEVIN. You have had a consensus on every compliance decision?

Dr. STANSBERRY. Every single one.

Senator LEVIN. The Compliance Review Group that you Chair has made over 100 certifications, as I understand it, for tests or programs as being compliant with the ABM Treaty, is that accurate?

Dr. STANSBERRY. Yes, sir.

Senator LEVIN. Over what period of time would that be?

Dr. STANSBERRY. I went back and counted that number. That is 100 since the advent of the SDI Program in about 1984.

Senator LEVIN. So in the last maybe 13 years?

Dr. STANSBERRY. Yes, sir. And I believe the number is 107 at the moment.

Senator LEVIN. And climbing. What is the relationship between your group and the Standing Consultative Commission delegation?

Dr. STANSBERRY. The delegation to the Standing Consultative Commission (SCC) is the U.S. Government delegation that discusses Treaty implementation issues with the other side in the Treaty. The SCC delegation is an Executive Branch-wide delegation, so it includes representatives from the Defense Department, the State Department, the Arms Control and Disarmament Agency, and the Intelligence Community. I am generally aware of what is going on because what they do can have some effect on my responsibility. As a matter of fact, about 16 or 18 years ago, I participated in a couple of sessions with that group. I do not directly participate anymore.

Senator LEVIN. You are assuring this Subcommittee that except for that 1 year period where there was an understanding relative to the software, which turned out to be inaccurate for whatever reason we could learn in a closed session, that the United States has not reduced the capability of any theater missile defense system in order to make them compliant with ABM, is that correct?

Dr. STANSBERRY. That is correct. That applies to all the other systems and in fact——

Senator LEVIN. All the other systems?

Dr. STANSBERRY. The Patriot, the Navy Theater Wide System, the Navy Area System are all theater missile defense (TMD) systems that are underway and that this statement applies to. With respect to THAAD, what happened was that THAAD did not need that capability through its demonstration and validation program and so it never actually limited the necessary capability. What it did was cost, as the Chairman said, I believe about $3.2 million dollars to assure during that period of time that we did not inappropriately develop the software. It did not result in any capability limitation of THAAD or any delay of THAAD, however.

Senator LEVIN. Does the United States and Russia each make their own National compliance determinations?

Dr. STANSBERRY. We do and it appears that the Russians do. They have not brought any to us asking for our approval.

Senator LEVIN. Are you familiar with the provision of the 1996 Defense Authorization Act which defines the ABM qualifying flight test as a flight test against the ballistic missile which in that flight test exceeds a range of 3,500 kilometers or a velocity of 5 kilometers per second?

Dr. STANSBERRY. Yes, sir, I am.

Senator LEVIN. Is that the test which is also sought to be agreed upon in our negotiations with the Russians? Is that the demarcation line?

Dr. STANSBERRY. That is the demarcation line. The characteristics for targets against which we would test theater missile defense systems under the ongoing demarcation negotiations.

Senator LEVIN. But that is the position that we have taken for the target demarcation line, is that correct?

Dr. STANSBERRY. Yes.

Senator LEVIN. And that is the same demarcation line as Congress has put into its, I think it was the Sense of the Congress language of 1996?

Dr. STANSBERRY. Yes, Senator.

Senator COCHRAN. Dr. Stansberry, our staff has prepared a chart showing the compliance status of U.S. theater missile defense systems as they stand today and as they would under the Helsinki demarcation agreement. If this chart is accurate, it seems like the Helsinki agreement does not help us much in determining compliance. In fact, with the exception of making clear that lower velocity systems can use data from the space-based SMTS sensor, the net effect seems to be to ban systems whose compliance status is uncertain at this time. I invite you to take a look at that chart. One is on display on the easel and you have been given a small copy there for your review.

[The chart referred to follows:]

Compliance Status of TMD Systems With and Without the Helsinki Demarcation Agreement

System	Status Without Demarcation Agreement	Status With Demarcation Agreement
PAC–3	Compliant	Compliant
Navy Area	Compliant	Compliant
THAAD	Compliant	Compliant
Navy Theater ("baseline")	Compliant	Compliant
Navy Theater (improved radar)	?	?
Navy Theater (launch on composite data)	?	?
Navy Theater (use of SMTS data)	?	?
Airborne Laser	?	?
Space-based Laser	?	Non-complaint
Space-based Interceptor	?	Non-complaint

Senator COCHRAN. The suggestion this chart makes is that those systems that you have already described that have already been approved—the advanced Patriot system, in the Navy Area, THAAD and Navy Theater have all been considered compliant by your group. Even without a demarcation agreement in accordance with the Helsinki statement, they would be compliant. Is that not correct?

Dr. STANSBERRY. That is correct.

Senator COCHRAN. And then these other systems are mentioned: Navy theater without an improved radar, launch on composite data, and use of SMTS data. Tell us what that is.

Dr. STANSBERRY. That is the Space Missile Tracking System. It more recently is called the low altitude component of the Space-Based Infrared System (SBIRs). Years ago, 2, 3 or 4 years ago, it was known as "Brilliant Eyes."

Senator COCHRAN. Right. SMTS is more politically correct than Brilliant Eyes I think is what they decided. And then the airborne laser is listed in the same category. Now, all of those we have put questions marks by is because their status is in question without a demarcation agreement and with an agreement based on the Helsinki statement would still be uncertain. And for the last two—the space-based laser and space-based interceptor—under the column "Status Without Demarcation Agreement" their compliance is in question, but with the Helsinki statement both would be non-compliant.

Do you agree with that depiction on this chart of those conclusions and if you do, tell us why you do and if you do not, tell us why you do not.

Dr. STANSBERRY. Most of them, Mr. Chairman. The three entries involving Navy Theater Wide with various levels of upgrade and the airborne laser are all uncertain today because there is no well-defined program to do either of those that would allow us to assess the compliance.

Senator COCHRAN. You mean any?

Dr. STANSBERRY. I'm sorry. Any of those, yes. The various upgrades to Navy Theater Wide are just simply too immature to understand the performance of the system well enough to be able to apply the process I discussed before. The airborne laser, an Air Force program, is in a similar state where it is not sufficiently mature to allow us to make compliance judgments. Our current understanding is that that program would probably not carry out the kind of test that would raise ABM Treaty issues for about another 5 years.

Now, where I guess I would take a certain exception to the chart is with respect to space-based lasers and space-based interceptors without a demarcation agreement. In the preparations for the Helsinki summit, the administration—well, let me back up for a second. The ABM Treaty contains an unequivocal ban on ABM systems and ABM components that are space-based. The question arose, "Can you have a theater missile defense system that is space-based, consistent with the ban on ABM systems, consistent with the understanding that it must not have capability to counter, must not be tested in an ABM mode?"

In the preparations for the Helsinki summit, the administration considered that question and, for a variety of technical reasons, concluded that it was probably impossible to have a space-based theater missile defense system which was not also a space-based ABM system and it was on that basis that the administration—the President—agreed at the Helsinki summit to ban space-based TMD, both interceptors and systems based on other physical principles.

Senator COCHRAN. Was that a decision made in consultation with Russian counterparts, or just here within the United States?

Dr. STANSBERRY. I believe that was a unilateral U.S. decision.

Senator COCHRAN. Was there any reason to suspect that the Russians would have lodged a complaint or protest if we had proceeded to use a space-based component in a theater missile defense system?

Dr. STANSBERRY. Well, based on past Russian performance, I would suspect that it would, almost certainly. They would raise a compliance issue.

Senator COCHRAN. Is there any way to have an early detection of a launch from a missile site aimed at us if you do not have a space-based detection system?

Dr. STANSBERRY. We currently have a ballistic missile early warning system. It is a sort of evolution of the system that goes back to the early 1970's, the Defense Support Program, and it provides notification of the launch within tens of seconds after it occurs.

Senator COCHRAN. And that is not prohibited under the ABM Treaty, right?

Dr. STANSBERRY. That is correct.

Senator COCHRAN. And what is prohibited, the ability to track?

Dr. STANSBERRY. No—Article VI(a) prohibits giving components the capability to counter and testing them in ABM mode. In Article V, it prohibits development, testing and deployment of space-based ABM systems; and we have defined systems as those which have in effect capability to counter as ABM systems in the past.

Senator COCHRAN. According to your testimony, we perform a comprehensive review of our systems every year to be sure they are compliant, but according to ACDA, which is responsible for evaluating Russian compliance with the Treaty, the U.S. does not systematically review Russian capabilities to ensure they are compliant; instead I am advised that we wait for evidence of non-compliance to appear and then we investigate that. But even then, if we do not turn up clear evidence of non-compliance, the Russians would not be charged with a violation.

Is this a double standard by which the burden of proof on the U.S. is to demonstrate conclusively to itself that any system we have could not possibly, even on a theoretical basis, have ABM capability, while with respect to Russian systems, the proof is related to a demonstration that they do have capability?

Dr. STANSBERRY. If there is a double standard here, Mr. Chairman, it is with respect to the timing of the issues, much less than the substance of the performance. Timing in the sense that we generally do not know what the Russians are doing until they have actually demonstrated something in a test. For our purposes, we would not pay the money to carry out a test that would be a violation. So we stop activity which might be inconsistent with the Treaty substantially earlier in the natural development of a program when we might see the comparable activity that the Russians undertake.

Now, with respect to the substance, I think basically we are consistent. As I have mentioned in my prepared statement, we have been concerned for a long time, going back into the early and mid—1970's about the potential ABM capability of Soviet air defenses. That concern was based on, as I would characterize today, hypothetical capability. We never saw the Soviet air defenses tested to intercept a strategic ballistic missile. So in that sense, again, the substance I believe is consistent, the timing may be different.

Senator COCHRAN. In your opinion, do any of our TMD systems or the potential or planned upgrades that are being considered have a militarily significant ABM capability? Specifically, do you think that the Navy Theater Wide equipped with a Cooperative Engagement Capability would have militarily significant ABM capability that would affect compliance decisions?

Dr. STANSBERRY. My honest answer to that is I do not know at the moment. The Navy has not proposed a concrete application of the Cooperative Engagement Capability for use with Navy Theater Wide so that we could assess how it actually impacts on ABM Treaty issues.

Senator COCHRAN. What about a more powerful radar for the Navy Theater Wide System, would that invalidate the compliance judgment that was rendered earlier?

Dr. STANSBERRY. The compliance judgment is based on the currently planned program. If the program changes we would have to review the compliance, so yes it would. A different radar would invalidate the existing compliance determination.

Senator COCHRAN. And what about the ability to launch its interceptor missiles based on information from external sensors, would that change the compliance judgment for the Navy Theater Wide?

Dr. STANSBERRY. That would also, for the same reason. They currently do not plan to do that and we only evaluated the currently planned program.

Senator COCHRAN. Back to the hypothetical, you do not deal in hypothetical questions?

Dr. STANSBERRY. Well, we have found it counter productive in the sense, and I am sure your legal background will be able to appreciate this, counter productive to try to make decisions in the abstract.

Senator COCHRAN. Like the Supreme Court. They do not like to——

Dr. STANSBERRY. I hesitate to compare what I do to what they do, but in that sense, yes.

Senator COCHRAN. Well, we are all supposed to be on the same side anyway. But if the Navy Theater Wide was able to launch its missiles based solely on data from the Space and Missile Tracking System, we mentioned that a while ago, the SMTS data, would that invalidate the previous compliance decision?

Dr. STANSBERRY. It would invalidate it. Whether it was OK under the treaty would depend on details that we do not have yet.

Senator COCHRAN. I have a few more specific questions, but I am going to yield at this point to Senator Levin for any questions he might have.

Senator LEVIN. Just a couple of more questions, Mr. Chairman.

I want to go back to that chart. As I understand your testimony, under Article V of the ABM Treaty, each party undertakes not to develop, test or deploy ABM systems or components which are sea-based, air-based, space-based which are the words used there or mobile land-based, is that correct?

Dr. STANSBERRY. That is correct.

Senator LEVIN. That is Article V?

Dr. STANSBERRY. That is Article V; yes, sir.

Senator LEVIN. In your judgment when you looked at that analysis prior to Helsinki, you reached a conclusion that those systems were banned by the ABM Treaty, is that correct?

Dr. STANSBERRY. The DOD compliance review process did not, it was the administration as a whole that made that judgment, but it was essentially what you say, that any TMD system—space-based TMD system that actually had TMD capability—would also have ABM capability and, hence, be an ABM system.

Senator LEVIN. Did you participate in that decision?

Dr. STANSBERRY. I did not.

Senator LEVIN. Did the DOD?

Dr. STANSBERRY. I do not know, sir.

Senator LEVIN. Would you get us, for the record, and tell us who was involved in reaching that?

Dr. STANSBERRY. I can do that, yes.

Senator LEVIN. If each party undertakes not to develop test or to deploy an ABM component which is space-based, is that not opprobrium or am I missing something here? Is it not by definition a violation of the ABM? Is there a way it does not violate the ABM Treaty to have a space-based laser or space-based interceptor?

Dr. STANSBERRY. Only if it did not have ABM capability. As I say, the judgment in the run up to the Helsinki summit was that, just as a practical matter, it was not possible to build a TMD laser that did not have ABM——

Senator LEVIN. Space-based.

Dr. STANSBERRY. Space-based, I'm sorry.

Senator LEVIN. That did not have ABM capability?

Dr. STANSBERRY. That is correct.

Senator LEVIN. If that is correct, that status would be non-compliant without a demarcation agreement, is that not correct?

Dr. STANSBERRY. That was the point that I made when I said I would take some exception to the chart. On that basis, it is not possible to make a TMD space-based system that is not also an ABM system. The two red question marks would be non-compliant.

Senator LEVIN. That is what I want to just clarify. Is that your judgment?

Dr. STANSBERRY. Yes.

Senator LEVIN. Other than that, do you have any disagreement with the chart?

Dr. STANSBERRY. No, I do not think so.

Senator LEVIN. In every other case it is the same with or without a demarcation agreement, right?

Dr. STANSBERRY. Yes.

Senator LEVIN. It is only in those two cases where the chart shows a difference and you disagree with the chart?

Dr. STANSBERRY. Yes.

Senator LEVIN. That is all I have. Thanks.

Senator COCHRAN. Dr. Stansberry, was the ABM capability attributed to THAAD in 1994 deemed militarily significant? If not, then why did the U.S. spend millions of dollars, $3.2 as you said, taking the cueing capability out of that system and then putting it back in?

Dr. STANSBERRY. Yes, Mr. Chairman, it was deemed militarily significant. It was deemed that it passed the threshold set up for Article VI(a).

Senator COCHRAN. In the National Missile Defense area, on its face, the Treaty seems to say that any system capable of protecting all of the U.S., even from a limited long-range ballistic missile threat, is prohibited.

How do we ever get around that in developing—or how can you ever decide that a National Missile Defense System is Treaty compliant?

Dr. STANSBERRY. I think the best way to characterize it is as sort of open to discussion at the moment. There has been no need to decide at the moment whether we believe that a National Missile Defense System is consistent or inconsistent with it. The Administration has already said that if we plan to deploy a National Missile Defense System and if that system is inconsistent with the Treaty, we would move to change the Treaty, not the system.

Senator COCHRAN. We had a meeting with our Subcommittee members and staff to become acquainted with what plans are underway for developing a National Missile Defense system and we learned that a request for proposals for a Lead System Integrator contract asked the contractors to consider a number of potential NMD architectures, including options to deploy x-band radars forward of the ABM deployment area. I know that none of these plans are firm yet, but let's assume that meeting a threat from North Korea, as an example, required a forward-based radar on the west coast of the United States. The Treaty defines an ABM radar as one constructed and deployed for an ABM role. If deployment of a forward based radar was necessary to complete intercepts of ICBMs, would not that radar meet the definition of an ABM radar in Article II?

Dr. STANSBERRY. It would certainly raise questions and it would depend upon the detailed nature of the radar and whether and what its other roles might be. That is an example of the kind of system architecture that might require us to seek modification of the Treaty.

Senator COCHRAN. Article III requires ABM radars to be located in the ABM deployment area or at test ranges. There again is a possible conflict with a forward-deployed radar that we might need, but which might not be Treaty compliant, is that not correct?

Dr. STANSBERRY. If that forward-deployed radar were deemed an ABM component, it clearly would be outside the allowed deployment area, yes.

Senator COCHRAN. What about a forward-deployed radar placed in South Korea or Japan?

Dr. STANSBERRY. The same comment, that if it were an ABM radar, it would clearly be outside the deployment area. Whether it is an ABM radar and whether it meets the requirements for an ABM component would be based on answers that are not available at the moment simply because the details are not here.

Senator COCHRAN. At this meeting in April that we had, General Cosumano said that if forward-based x-band radars were required, they will likely be less powerful versions of the ground-based radar, which is the NMD system's ABM radar. If this were the case, would not that radar be of the same type as the radar tested in an ABM mode and therefore an ABM radar under Article II's definition?

Dr. STANSBERRY. Not necessarily, because the type of rules set out in Article II carry some implications about sort of being identical and if it is a similar radar, but of a smaller size, that may mean it is a different type. Again, we cannot make judgments on questions like that until we see the actual system design of the particular characteristics of the radar.

Senator COCHRAN. The suggestions about the space-based components bothers me to some extent. I am going to read a statement made by Mr. Bob Bell in March of 1997. He is of course from the National Security Council staff. I am going to quote this from a press conference statement:

"When we looked at this, both as a matter of compliance law and as a matter of technological assessment, we determined that in fact it would be impossible to distinguish between an orbiting laser and orbiting battle station armed with kinetic kill missiles that one side would claim was only designed to intercept theater range missiles coming through space and somehow not have the capability to destroy strategic missiles going through space."

Are you familiar with that assessment and did you participate in that decision or that conclusion about the space-based TMD system?

Dr. STANSBERRY. That is the same assessment I was discussing with Senator Levin as part of the run-up to Helsinki. I am familiar with it and no, I did not participate. It is the thing that leads to dispute about the two red question marks on the chart.

Senator COCHRAN. He also said the following: "After all, if you have a laser in space that has the power to burn a hole through the side of a missile and blow it up, it can do that whether the missile was an ICBM or an intermediate range missile. So from our perspective, there was no demarcation possible between space-based TMDs and space-based ABMs and we see this agreement as a logical corollary of the existing prohibition in the Treaty."

Is this the position of the review group on this subject?

Dr. STANSBERRY. The Compliance Review Group has not addressed that specific question and so I do not have any particular position on it. I can understand that for a number of technical reasons it is probably easier to kill, to destroy strategic missiles from space than it is to destroy theater missiles from space. The strategic missiles get closer to the laser or the interceptor and hence, are just easier to destroy. It is on the basis of considerations like that that any TMD system that is space-based will probably have capability to destroy strategic systems as well, and hence, be a violation of the Article V ban on space-based ABM.

Senator COCHRAN. What would be the difference then between that and an airborne laser? Would it not be able to burn a hole in a missile and kill it just as well whether it is a strategic missile or a theater missile?

Dr. STANSBERRY. It would, depending on where the airborne platform is. The space-based platform for a space-based laser is subject to the laws of physics determining satellite orbits. An airborne platform is subject to some substantially different considerations and I think that those differences would probably come into play in evaluating an airborne laser compliance.

Senator COCHRAN. But they are both prohibited in Article V. It is the same article prohibiting space-based systems as prohibits air-based systems.

Dr. STANSBERRY. That is correct—space-based and air-based ABM; and it is important to understand the distinction here. The assessment prior to Helsinki was that a space-based interceptor or a space-based laser for TMD would likely have ABM capability.

That is not the conclusion for an airborne laser. While we have not had to look at the airborne laser in any detail, I suspect that we can come to a different conclusion. The use of an airborne laser for theater missile defense is not necessarily an airborne laser contrary to the ABM Treaty.

Senator COCHRAN. It sounds in conclusion that, first of all we have rules that we are applying against ourselves that may make our costs of developing theater missile defense systems, even if they are Treaty compliant, more expensive than they would otherwise be if we were not constrained by our own interpretations of the ABM Treaty. Is that not correct? Is that not borne out by the record?

Dr. STANSBERRY. We have at least the one instance of THAAD where it has been more costly to abide by the Treaty than if were there no Treaty.

Senator COCHRAN. My next question then is do you not agree that our interpretations of the ABM Treaty constrain us to the extent that we have no plans or proposals to deploy theater missile defense systems that are sophisticated enough to fully protect us in the case of missile attack? In other words, we are not preparing and not planning to deploy the systems that are so advanced that they guarantee us a higher state of security because of our own interpretation of the ABM Treaty, is that not correct?

Dr. STANSBERRY. I do not think so, Mr. Chairman. The process by which we define and approve programs starts with requirements. The Theater Missile Defense programs that are currently underway have their requirements defined in that process. These are requirements that are not limited by the ABM Treaty and all of the systems that we are currently developing and deploying meet the requirements.

Senator COCHRAN. The requirement is judged on the basis of the threat that you perceive exists, is that correct?

Dr. STANSBERRY. That is correct.

Senator COCHRAN. Well, I know that to assess the threat, we end up having to be in a closed session to really find out the details about what the emerging threats are, so we cannot get into that part and we do not purport to say that that is a part of this discussion. We are looking at our compliance review process and I think you have been very helpful to us in understanding how it works and what the standards are. It seems to me that the standard we have may be subject to question.

I am going to ask you what your assessment of it is. If you were the only person and you did not have to develop a consensus, would you say that the standard that we have now is the best standard or the standard that we ought to have in place or how would you improve it if you could do that?

Dr. STANSBERRY. I think the standard works as we have applied it. We have not limited the performance capability of any TMD system that we have otherwise wanted to develop and test. On that basis, I think the standard is working. I do not believe we have gone as far as we can go because there are additional questions about additional performance based on potential improvements and the chart that the staff put together is an example. There are potential improvements for the Navy Theater Wide System that raise some questions. We have not evaluated those questions, but I think it is fair to evaluate them in the context of the same compliance standards and practices that we have used in the past.

Senator COCHRAN. Well, why then is it necessary for us to reach an agreement based on the Helsinki statement? All of our systems that we have in place now or that we have planned are compliant.

Dr. STANSBERRY. I think there are two or three reasons. One is that the question marks on the chart are question marks. We do not know what we might want to do in the future and a demarcation agreement would take away the question marks. Another element is that it would, and I think this is a substantial sort of political consideration, it would remove as potential item of dispute between us and the Russians this area of the ABM Treaty. As you mentioned in your opening statement, the connection between the ABM Treaty and strategic offensive limitations with the Russians is important here, so that if we have an element of dispute between the Russians, it has the potential to undercut strategic offensive limitations, such as START, START II, and maybe a future START III.

Senator COCHRAN. Thank you very much. Senator Levin?

Senator LEVIN. Just one additional question. There has been long-standing differences in the Congress over the ABM Treaty and its value and its impact on arms control and arms reduction and so forth, but where I think there has been a consensus is on theater missile defense systems. I think all of us want to develop and to deploy effective TMD systems.

Can we develop and deploy effective TMD systems and still comply with the ABM Treaty in your judgment?

Dr. STANSBERRY. I believe so, Senator Levin. The systems that we have currently under development, TMD systems, have been designed to meet the approved TMD requirements and we can do that consistent with the existing ABM Treaty. Whether that continues to hold in the future with additional potential threats is as you know, a future question.

Senator LEVIN. Thank you. Thank you, Mr. Chairman.

Senator COCHRAN. Thank you, Senator.

Dr. Stansberry, thank you very much for your assistance and your testimony today at this hearing. We will continue our series of hearings on subjects under the jurisdiction of this Subcommittee. We thank all of the staff who have worked so hard to help make these such a great success. The Subcommittee is adjourned.

[Whereupon, at 3:44 p.m., the Subcommittee was adjourned.]

APPENDIX

QUESTIONS FOR THE RECORD

Question No. 1

DEMARCATION STANDARD OF THE HELSINKI PRINCIPLES

Question: At their Helsinki summit in March, Presidents Clinton and Yeltsin announced a set of elements for a demarcation agreement. If that agreement is ever codified—and in June the Standing Consultative Commission adjourned once again without finalizing that agreement—would an agreement based on the Helsinki principles set an unambiguous demarcation standard for your use in making compliance decisions for higher velocity systems or for other advanced systems such as the airborne laser?

Answer: The agreement for higher velocity theater missile defense (TMD) systems based on the elements agreed by the Presidents at Helsinki does not establish a definitive demarcation standard. Compliance assessment for higher velocity theater missile defense systems and other advanced systems such as the airborne laser would remain a national responsibility. The agreement for lower velocity TMD systems does establish a definitive demarcation standard.

Question No. 2

CLARITY OF "DEMONSTRATED CAPABILITY" STANDARD

Question: The Clinton Administration had proposed a "demonstrated capability" standard that would deem any TMD system compliant as long as it was not tested against a target missile exceeding 5 km/s or 3500 range during TMD testing. Would such a standard give you the clarity you need to make unambiguous compliance judgments, and is this an adequate standard to ensure TMD systems do not have significant ABM capability?

Answer: Adoption of such a standard, as agreed in the draft demarcation agreement on lower velocity systems, would certainly facilitate our compliance determinations. In the U.S. view, such a standard, together with various confidence building measures, would be adequate to ensure theater missile defense (TMD) systems do not have significant ABM capability, but we believe it is important that all ABM Treaty parties agree on such a standard. However, without the benefit of the "demonstrated capability" standard, we have already certified as treaty compliant all existing U.S. TMD programs that are sufficiently mature to allow a compliance assessment, including the higher velocity Navy Theater Wide system.

Question No. 3

NUMBERS OF MISSILES IN COMPLIANCE CALCULATIONS

Question: Do the actual numbers of offensive and defensive missiles—not types but the actual quantities—enter into your compliance calculations? Should these numbers be part of those calculations?

Answer: The Department of Defense has not used the actual numbers of offensive and defensive missiles as a consideration for compliance assessments. We have made such assessments on the basis of estimating the ability of a single TMD interceptor and a single TMD radar to counter a single incoming strategic reentry vehicle. Using this approach we have been able to certify as Treaty compliant all TMD systems that are sufficiently mature to make evaluation possible. The Administration has made no determination regarding whether the use of the numbers of offensive and defensive missiles in a compliance assessment would be appropriate or how that would be done.

Question No. 4

"FORCE-ON-FORCE" COMPLIANCE CONSIDERATIONS

Question: In their May 10, 1995 joint Summit Statement, Presidents Clinton and Yeltsin declared that "TMD systems may be deployed by each side . . . will not pose a realistic threat to the strategic nuclear force of the other side. . . ." This suggests that your compliance determinations should be judged on the basis of whether a TMD system could negate a large scale strategic attack and not just have some theoretical capability against a single reentry vehicle under "perfect" conditions.

- Are such "force-on-force" calculations the basis for your compliance determinations? If not, why not?
- Have Defense Department legal authorities issued any opinions on whether such a standard might be used? If so, please provide that opinion for the record, as well as the written request to the OSD General Counsel which generated this legal opinion.
- If so, is this legal opinion the basis for your current compliance standards?

Answer:

- No. The Helsinki Summit principles served as the political basis for the demarcation agreements. The principles themselves have not been agreed to be legal standards for compliance. All of the principles, including force-on-force, served as the basis for negotiators in reaching agreements and will serve as the basis for paying increased attention to practical and realistic assessments in implementing the Treaty.
- Yes, one opinion. For reasons of confidentiality, we do not believe it would be appropriate to provide for the record the opinion or the written request.
- No.

Question No. 5

"TRIVIAL" PORTION OF THE OTHER SIDE'S MISSILES

Question: The unclassified section of the October 31, 1996, Report to Congress on ABM Treaty Compliance Certifications of Theater Missile Defense Systems states: "A non- ABM system under consideration would only be inconsistent with the obligations if it had capabilities to counter strategic ballistic missiles of types that comprise more than a trivial portion of the other side's strategic ballistic missiles reasonably projected to be in existence when the system is to be deployed."

- How is the term "trivial" defined? Is this a legal definition, or one based on assessments of militarily relevant capabilities? How is "non-trivial" different from "militarily significant"?
- This report refers to capability against "types" of strategic ballistic missiles, not numbers of missiles.
  - This logic implies that if a TMD system were determined to have some limited theoretical capability to intercept strategic missiles—for example, the same capability THAAD was determined to have in 1994—and the other side had only one type of ballistic missile, then that TMD system would be deemed to have capability against 100% of the opposing side's strategic ballistic missile force. Is this what is implied by the report?
  - Since the number of actual missiles is not considered in compliance determinations, if the United States had only one of the type of TMD interceptor in the above example, and the other side had 10,000 of their one type of strategic missile, would the TMD system then be deemed to have capability to counter the other side's entire force?

Answer:

- In this case, trivial is meant to have its ordinary meaning (e.g., of little significance) and has no special legal significance. It has not been specifically quantified. In considering THAAD compliance in 1994, it was noted that certain Russian strategic ballistic missiles, which are relatively easy to counter, comprise approximately one and one-half percent of the projected Russian strategic force. This was characterized as a trivial portion of that force. No effort was made to relate "non-trivial" and "military significance."
- No. What is meant is that one TMD interceptor and radar would have capability against one of that type of ballistic missile (i.e., one reentry vehicle).
- No, for the reasons provided above.

Question No. 6

FORWARD-BASED X-BAND RADAR

Question: You stated in your testimony that whether a forward-based radar for use in national missile defense was an ABM radar "would depend upon the detailed nature of the radar and whether and what its other roles might be." Please explain how "other roles" served by a forward-based X-band radar which was constructed and deployed explicitly as part of the NMD architecture could prevent it from being deemed an ABM radar in accordance with the definition of that term contained in Article II of the ABM Treaty.

Answer: We have not found it useful or appropriate to try to address treaty compliance questions in the abstract, but rather we address them in the context of information about specific and detailed plans. Questions regarding the use of a forward-based radar for national missile defense and other purposes fall into this category.

Question No. 7

THRESHOLD FOR MILITARILY SIGNIFICANT ABM CAPABILITY

Question: Your written testimony contains a statement regarding "military significance" but it quotes a passage from a 1986 report to Congress on *Soviet* compliance, not U.S. compliance. "Military significance" is part of U.S. evaluation of Russian TMD systems, but is it, and has it been since 1993, an *explicit* part of the Compliance Review Group's assessment of *U.S.* TMD systems? If so, how is "military significance" quantified? What is the threshold above which TMD systems are deemed to have militarily significant ABM capabilities?

Answer: Military significance is taken into account in compliance assessments of both U.S. and Russian TMD systems. However, the term "military significance" is not explicitly defined; it has not been applied as a standard for compliance, and there is no defined "threshold." Rather, it is taken into account inasmuch as military considerations, such as the nature of the opposing strategic ballistic missiles and the simulated, hypothetical capability of the TMD system against those missiles, are considered in compliance determinations. Judgments about compliance are made on a case-by-case basis in light of the particular circumstance of each case.

Question No. 8

JOINT STAFF ASSESSMENT OF MILITARILY SIGNIFICANT ABM CAPABILITY

Question: According to a briefing presented to subcommittee staff on June 23, 1997, by officials of the Arms Control and Disarmament Agency, the Joint Staff is responsible for determining whether Russian TMD systems have militarily significant ABM capability. When was the last time the Joint Staff made such an assessment? Does it also make these assessments for U.S. systems? If not, does the CRG use the same analytical model the Joint Staff employs?

Answer: The Joint Staff participates fully in the interagency assessments of Russian TMD systems as well as DOD assessments of U.S. TMD systems, providing military advice for both. Rather than making assessments of "military significance" per se, the Joint Staff provides its views on military considerations such as the nature of the opposing strategic ballistic missiles and the capability of the TMD systems against those missiles. Neither the Joint Staff nor the DOD Compliance Review Group has adopted an "analytical model" specifically for evaluation of "military significance."

Question No. 9

MILITARY SIGNIFICANCE OF THAAD'S ABM CAPABILITY

Question: You stated in your testimony that, in 1994, THAAD's theoretical ABM capability was deemed "militarily significant." That statement directly contradicts what was reported to Congress in the January 1994 report on THAAD compliance, as well as a response for the record from BMDO Director O'Neill in testimony before the SASC on May 11, 1994, in which be stated, "This 'theoretical' ABM capability, however, would not be militarily significant in light of system limitations and operational considerations. That is, in real-world scenarios, THAAD could not perform its defensive mission against strategic-class missiles, even in limited engagements. . . . In sum, THAAD simply would not suffice as an ABM system; it would be easily overwhelmed by the Russian strategic missile force." Does your testimony change the Administration's position as previously reported to Congress, or was your testimony incorrect?

Answer: My response to the Chairman's question on this point was incorrect. What 1 should have said was that military significance was taken into account in the evaluation of THAAD. As noted above, the term "military significance" is not defined and has not been applied, as such, as a standard for compliance.

Question No. 10

ONE-ON-ONE ANALYSIS TO MEASURE MILITARY SIGNIFICANCE

Question: Your testimony states that you assess the "military significance" of U.S. TMD systems on the basis of "the simulated, hypothetical performance of a single TMD interceptor missile and radar to intercept a single reentry vehicle from certain Russian [strategic missiles]." But according to General O'Neill's testimony, the THAAD system was declared non-compliant using this measure even though it had no militarily significant operational capability. Given his testimony, how can the one-on-one analysis you described be considered a measure of military significance?

Answer: As noted above, military significance is taken into account in compliance assessments. However, "military significance" is not explicitly defined and has not been applied as a standard for compliance. Rather, military significance is taken into account inasmuch as military factors such as the nature of the opposing ballistic missiles and the capability of the TMD system against those missiles are considered in compliance determinations. The conclusion that General O'Neill described in his testimony resulted from including certain operational considerations, such as numbers and locations of deployments, that went beyond the one-on-one analysis used in the compliance assessment for THAAD. Subsequent to General O'Neill's testimony, and using our current approach, THAAD was determined to comply with the ABM Treaty.

Question No. 11

"MILITARY SIGNIFICANCE" AS A STANDARD FOR COMPLIANCE

Question: Since THAAD's capability was deemed not militarily significant in 1994, but THAAD was still determined to be non-compliant, how is it possible to state that "military significance" is the standard by which the CRG judges the compliance of U.S. systems?

Answer: As noted above, "military significance" has not been used as a standard for judging compliance of U.S. systems.

Question No. 12

"MILITARY SIGNIFICANCE" AS A STANDARD FOR COMPLIANCE

Question: You stated in your testimony that you do not take into account operational factors of TMD systems in making compliance assessments. On what basis did you state that "military significance" is a compliance standard if operational military factors are not considered?

Answer: As noted above, "military significance" has not been used as a standard for judging compliance of U.S. systems.

Question No. 13

JOINT STAFF CONCERNS

Question: You stated that all compliance determinations have been consensus decisions which included the concurrence of the Joint Staff. Has the Joint Staff made known to you any concerns on its part that current compliance methodologies, assumptions, or standards are inadequate and should be replaced with ones that better measure the true military significance of the capability of TMD systems? If so, what are the nature of those concerns?

Answer: The Department of Defense will continue to ensure that DOD programs fully comply with all treaty obligations, but it also will continue to refine compliance methodologies and assumptions to ensure that these programs are not unnecessarily constrained. Our compliance methodology has evolved to incorporate additional relevant considerations. The Joint Staff has participated fully and constructively in this process.

Question No. 14

ONE-ON-ONE ENGAGEMENTS AS A MEASURE OF MILITARY SIGNIFICANCE

Question: Please explain how the "simulated, hypothetical one-on-one engagements" you mentioned in your testimony represent a reasonable operational scenario from which a true measure of military significance can be derived?

Answer: As noted above, our one-on-one compliance methodology has not taken all operational considerations into account, and no claim is made that a "true measure of military significance" has been derived.

Question No. 15

NEGOTIATED OR UNILATERAL STANDARD

Question: Is the current standard by which TMD programs are judged for ABM Treaty compliance the result of a negotiated agreement with Russia or the U.S.S.R., or was the standard determined unilaterally by the United States?

Answer: Compliance judgments, which are derived from the provisions of the Treaty, have been made unilaterally on a case-by-case basis rather than on the basis of a particular standard for compliance. The methodologies used thus far to evaluate TMD systems for compliance are not the result of negotiated agreement.

Question No. 16

ASSESSMENT OF SPACE-BASED TMD

Question: Robert Bell of the National Security Council Staff has stated that prior to the Helsinki summit, the administration examined space-based lasers and- interceptors "as a matter of compliance law and technological assessment." Presumably the compliance expertise for this assessment would come from the executive branch agency responsible for making compliance judgments on U.S. systems, which by Defense Department directive is the DOD Compliance Review Group, with technological assessments from the missile defense experts in the Ballistic Missile Defense Organization. Yet you stated in your testimony that the CRG was not involved in this assessment and that you, as chairman of the CRG, did not know if the Defense Department was involved in the determination. Please provide the subcommittee the technological and "compliance law" assessments that formed the basis for this decision, as well as a list of the participants in those assessments.

Answer: The USG decision to agree at the Helsinki Summit to prohibit space-based TMD interceptor missiles and components based on other physical principles that are capable of substituting for such interceptor missiles was made, like other decisions on the negotiations, through the interagency process, in which all relevant agencies, including DOD (both OSD and the Joint Staff) participated at senior levels. Issues concerning space-based TMD systems, including technical and legal considerations, were discussed in appropriate detail during interagency deliberations, although no formal (i.e., written) technological or "compliance law" assessments were prepared. The Administration concluded that, as a practical matter, no demarcation between space-based ABM and space-based TMD systems was feasible. The USG has maintained the position that it could accept a ban on space-based TMD interceptor missiles since Spring of 1994.

Question No. 17

COMPLIANCE OF SPACE-BASED TMD SYSTEMS

Question: You stated in your testimony that you could not speculate on whether certain specific capabilities—for example, an upgraded Aegis radar, Navy Theater Wide aided by Cooperative Engagement Capability or launch on composite data, and forward-based X-band NMD radars—would be compliant because "there are no well-defined programs . . . that would allow us to assess the compliance" and because the capabilities "are just simply too immature to understand the performance of the system." Yet you said in your testimony that it was your judgment that "it's not possible to make a TMD space-based system that isn't also an ABM system," even though you had not participated in and were not privy to the analysis that led to this conclusion. Please explain how you can be certain that an undefined space-based TMD capability would necessarily be non-compliant when, according to your own testimony, any other definitive compliance determination requires "mature" and "well-defined" programs?

Answer: I did not mean to imply in my testimony that it was my judgment that an undefined space-based TMD capability would necessarily be non-compliant. Rather, I was attempting to explain the Administration's conclusion that, as a practical matter, no demarcation between space-based ABM and space-based TMD systems was feasible. That conclusion was not reached in the context of a compliance review of specific and detailed program plans, which are required for DOD compliance determinations.

Question No. 18

COMPLIANCE CONSIDERATIONS FOR AIRBORNE LASER TMD

Question: You stated the following in your testimony: that Article V of the ABM Treaty prohibits space-based ABM components and thus would also prohibit space-based TMD components which had ABM capability; that U.S. compliance assessments are made on the basis of "simulated, hypothetical performance" in one-on-one engagements; and that "operational factors including basing modes, the number of launch platforms and missiles, the location of TMD systems and the character of a likely strategic attack" were not part of your compliance evaluations. Yet you also stated that whether an airborne laser is compliant might depend "on where the airborne platform is" and that such a system would be "subject to some substantially different considerations." If the locations of TMD systems are not considered in your compliance determinations, why would "where the airborne platform is" matter in a compliance determination for the airborne laser? On what basis would "different considerations" be introduced for the airborne laser, what would they be, and why wouldn't the same considerations be used for determining compliance of all TMD systems?

Answer: In my testimony, I was speculating in response to a general, hypothetical question from the chairman, so my response should not be taken as a considered compliance assessment. With respect to the particular question, however, since we evaluate Th4D compliance on a case-by-case basis taking into account detailed characteristics of each system, the considerations relevant to compliance for an airborne laser would likely be different from those for systems with ground-based interceptors. These considerations would be evaluated when the system design is appropriately mature.

Question No. 19

SPACE-BASED AND AIRBORNE LASERS

Question: Article V of the ABM Treaty bans both space-based and air-based ABM systems, and you testified that a space-based TMD laser would "have capability to destroy strategic systems as well, and hence, be a violation of the Article V ban on space-based ABM." Please explain why the capability of a laser to, in the words of Robert Bell, "bum a hole through the side of a missile and blow it up," is sufficient to deem a space-based laser non-compliant but is not sufficient to make the same determination for an airborne laser.

Answer: The Administration concluded that, as a practical matter, no demarcation between space-based ABM and space-based TMD systems was feasible. Neither DOD nor the Administration as a whole has addressed the compliance question for the TMD airborne laser, and DOD has no plan to do so until the program is sufficiently well defined.

Question No. 20

ARTICLE BY LAWRENCE GOLDMUNTZ

Question: Did the analysis you said preceded the Helsinki agreement to ban space- based TMD consider contrary analysis which argues that space-based-TMD systems could be deployed without having ABM capability, such as that contained in Dr. Lawrence Goldmuntz's article "Poor Man's MRVs and Space Defense," in the Fall 1996 Strategic Review? If so, on what basis were these arguments rejected?

Answer: Arguments such as those in the article by Dr. Goldmuntz were considered, but were not accepted. The USG has maintained the position that it could accept a ban on space- based TMD interceptor missiles since Spring of 1994.

Question No. 21

INTERAGENCY LAWYERS

Question: You stated in your testimony that DOD has sole responsibility for making compliance determinations and DOD Directive 2060.1 confirms this. To what extent have lawyers from other executive agencies had a significant role in past compliance decisions? Do they have an official role and, if so, what is it and where is it prescribed in regulation or other administration policy statement? Has an interagency lawyers' group always been involved in these decisions to the extent it has since 1993? Who decides when to involve the interagency lawyers' group and based on what criteria?

Answer: As a general rule, lawyers from the other agencies do not get involved in DOD compliance assessments. Sometimes, however, the interagency legal community is consulted. The Department has the obligation to ensure that its activities comply with arms control agreements and takes that obligation very seriously. Accordingly, when certain significant issues of treaty interpretation have arisen, the office of the DOD General Counsel has consulted the interagency legal community in order to ensure that we proceed under correct treaty interpretations and that our activities will remain in compliance. Decisions to consult are made on a case-by-case basis by the Office of the DOD General Counsel, generally in consultation with representatives of the other offices represented in the Compliance Review Group. Interagency consultations have become more frequent in the 1990s as the issues have become more complex.

Question No. 22

RUSSIAN COMPLAINT ABOUT THAAD

Question: Have the Russians ever charged that THAAD was not compliant with the ABM Treaty?

Answer: No.

Question No. 23

RUSSIAN COMPLIANCE REVIEW

Question: Has the United States ever briefed or otherwise informed Russia of the results of its compliance decisions? Has Russia ever shared with the United -States any information on its compliance review process or the results of its reviews?

Answer: The United States has informed Russia of the results of certain compliance decisions. For example, during discussions in the Standing Consultative Commission we have briefly discussed thel996 decision on the THAAD system's compliance, and addressed the U.S. compliance process and the 1995 compliance report on the Navy Theater Wide system. There have been no instances of Russia providing comparable information to the United States regarding its compliance review process or the results of its reviews.

Question No. 24

INCREASED COST OF THAAD

Question: The October 31, 1996 Report to Congress on ABM Treaty Compliance Certifications of Theater Missile Defense Systems stated that the "current estimate" of the increased cost of the THAAD system due to the decision to withhold, and then reinstate, cueing software, was $3.4 million. This figure differs somewhat from the amount you cited in your testimony. What are the current and estimated final amounts-spent by the United States to keep THAAD from having a capability which was later determined to be compliant?

Answer: The number I quoted, $3.2M, was incorrect. The number that was reported in the October 31, 1996, Report to Congress is the correct number—$3.4M.

Question No. 25

REASONS FOR CHANGE IN THAAD COMPLIANCE ASSESSMENT

Question: Was the change in THAAD's compliance status between 1994 and 1996 due primarily to improved knowledge of the actual (vs. intended) technical capabilities gained through the demonstration/validation test program, or to changes in the assumptions used to make the compliance assessment?

Answer: As I noted in my testimony, the changes to THAAD were of a programmatic nature (i.e., related to system design). I also stated that the capabilities assumed for THAAD in early compliance deliberations were revised based on a better understanding of the THAAD performance when the CRG conducted its review of the program in 1996. When the CRG conducted the 1996 review, nothing had changed regarding our approach to determining ABM Treaty compliance. What did change was our understanding of the capabilities of THAAD as the system had matured and preliminary conservative performance assumptions were replaced by more accurate information.

○

APEX
TP
9780160558702
76901945 – 44